INGHAM CO. LIBRARY

J 574.1

Horsburgh, Peg

Living light / 4.95

000871 NOV 7 96

MA AUG 1984

Discarded by
Ingham County Library

INGHAM COUNTY LIBRARY
MASON BRANCH
145 W. ASH ST.
MASON, MI 48854

# LIVING LIGHT:
## Exploring Bioluminescence

# LIVING LIGHT:
## Exploring Bioluminescence

### Peg Horsburgh

JULIAN MESSNER      NEW YORK

Copyright © 1978 by Amelia M. Horsburgh
All rights reserved including the right of reproduction in
whole or in part in any form. Published by Julian Messner,
a Simon & Schuster Division of Gulf & Western
Corporation, Simon & Schuster Building, 1230 Avenue of
the Americas, New York, N.Y. 10020.

Manufactured in the United States of America

Design by Alex D'Amato

Drawings by Virginia Arnold

Library of Congress Cataloging in Publication Data

Horsburgh, Peg.
 Living light.

 Includes index.
 SUMMARY: Discusses the causes of bioluminescence,
describes animals and plants that exhibit this charac-
teristic, and relates the possible benefits of current
research in the field.
 1. Bioluminescence—Juvenile literature.
[1. Bioluminescence] I. Arnold, Virginia Ann.
II. Title.
QH641.H64      574.1'9125      78-1684
ISBN 0-671-32849-2

To my friends Wendy and Jack Murphy, and Lavinia Dobler.

The Murphy's summer home, on Blueberry Island, Maine, was where I first saw wood aglow. My interest in bioluminescence having been sparked by this eerie sight, I could not let the subject rest until this book was completed.

Lavinia Dobler has been my inspiration. She has continued to share her enthusiasm and knowledge about writing throughout the project.

# Acknowledgments

The author wishes to acknowledge with gratitude the help of Dr. Grace Picciolo, of the Goddard Space Flight Center, and Dr. Frank Johnson, of Princeton University, who provided her with much first-hand information about current experiments with bioluminescence, and with many photographs.

She wishes to thank Roy Pinney for making her trip to the Goddard Space Flight Center possible, and for sharing his expertise in photography and natural sciences.

Also, many thanks are extended to Virginia Arnold whose sensitive illustrations have added breadth to this book.

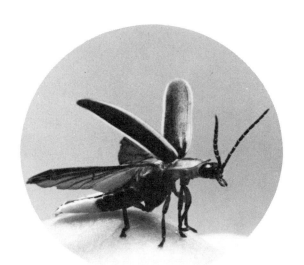

# CONTENTS

CHAPTER 1
    NATURE'S LIVING LIGHT 13
    2
    WHAT LIGHTS THE LIGHT? 18
    3
    FLASHING FIREFLY 27
    4
    SEA FIRE AND OTHER WATER LIGHTS 32
    5
    DEEP SEA LIGHTS 43
    6
    PLANT LUMINESCENCE 53
    7
    WHY LIGHT UP? 57
    8
    COLD LIGHT AND MEDICINE 67
    9
    COLD LIGHT AND TECHNOLOGY 75
    10
    HELPING SCIENCE 78
CHART 87
GLOSSARY 88
INDEX 91

# LIVING LIGHT:
## Exploring Bioluminescence

# CHAPTER I

# Nature's Living Light

The time was the year 1672. The place, a scientist's laboratory in England. Chemist Robert Boyle was writing a letter to The Royal Society, a group of well-known scientists of the day:

*Dear Sirs (he wrote),*

*I think the information in this letter will interest you. It tells of a strange experience I've had with that mysterious light in nature—bioluminescence.*

*Tonight my cook came running into the laboratory.*

*"There's ghosts down there, sir," she yelled. "In the cellar, sir, it's all lit up! No sir, I won't go back down there!"*

Robert Boyle worked in a laboratory much like this one in the late 1600's.

*The poor woman was so hysterical—what could I do but investigate?*

*It was true. The cellar was lit up, and the light did not come from a candle. It came from a dead chicken which had been hanging from a hook in the ceiling for several days. I had asked the cook to fix the bird for dinner, and there it was, aglow in a blue-green light.*

*I was astonished. As you all know, I have long had an interest in creatures which make their own light. I have written about my experiments with glowworms. But, gentlemen, I had never seen such a thing as this light. I decided I had to experiment with it immediately.*

*In short, here are my findings:*
- *The bird's light gave no heat.*
- *I poured wine on part of the chicken. The light dimmed where it was touched by the wine.*
- *The wind tonight was north by northeast, and traveled at about 15 miles an hour.*
- *I covered part of the chicken to keep air from it. Where there was no air, there was no light.*

- *A full moon rose at 8:00 P.M.*

*I want you to know yet another fact. When finished with my experiments, I asked the cook to prepare the chicken for dinner.*

*Perhaps you can imagine her horror, but the cook did as asked, and gentlemen, the bird was delicious!*

Robert Boyle was a pioneer in the world of research in bioluminescence. He never did learn it was bacteria which had made his chicken light up. *Bacteria* are plants that are so small they cannot be seen by the naked eye, and must be viewed through a microscope.

But Boyle was a true scientist and he recorded all the facts about his experiments. The wind and moon had nothing to do with the bird's light. But his findings on the lack of heat in the light, and the effect of air and water were important. With these facts, Robert Boyle set the stage for other scientists' work with bioluminescence over the centuries, work that continues today.

What exactly is bioluminescence? And why has it excited the imagination of people for hundreds of years?

*Bio* comes from a Greek word meaning *life*. *Luminescence* is from a Latin word which means *light. Bioluminescence* is living light found in nature.

Light is the result of some kind of energy—the energy of electricity, or fire, or the sun's energy. It takes energy to make bioluminescence, too. All bioluminescence comes from the energy released as plants and animals use their food. This is called *metabolic energy*.

Bioluminescence occurs in a wide variety of plants and animals. Some shrimp, squid, and clams luminesce. Sharks, flashlight fish, and jelly fish luminesce. Near the ocean's surface are tiny one-celled organisms which give off beautiful showers of light. Fireflies flash their lights, and glowworms glow.

Among plants, some bacteria and fungi luminesce. Mushrooms and toadstools may light up. Dead wood can glow. Meat, like the chicken Robert Boyle wrote about, can luminesce. Bioluminescence has been known to grow on people, too!

Living light comes in a variety of colors. It might be red, blue, green, yellow, white, or any combination of these colors.

The beauty of bioluminescence might be enough reason for people's interest in it. It is exciting to explore the ocean's depths and come upon a fish alive with light. Surely it is an eerie experience to walk along a beach and see light in the distance . . . and to find upon reaching the glow that it is a piece of driftwood.

But there's more to the story of bioluminescence than its fascination. Since Robert Boyle's time, scientists have learned a great deal about what causes luminescence. They are putting their findings to work in medicine and technology. Chemicals which make a firefly's tail light up are now used by doctors to diagnose some diseases. These same chemicals are used to detect water pollution.

Also, bioluminescence has one quality no other light has: it is cold light. None of the energy used to make bioluminescence is lost in the form of heat. Each time you turn on an electric light about 60% of the energy it takes to make the light is lost to heat. This is not so with bioluminescence. Scientists are looking for ways to make this efficient form of light in the laboratory. The energy crisis makes the search an important one.

And scientists will continue searching for ways they can use their knowledge of bioluminescence to help people. This search could become an important part of your future.

# CHAPTER 2

# What Lights the Light?

What lights the light of bioluminescence? Until recent years, the answers were just guesses, full of mystery and superstition.

Firefly lights were thought to be spirits of dead people. Ground-up fireflies were used to cure earaches. People believed dead tree trunks glowed because of magic.

Robert Boyle was the first scientist to find any true answers to the question. He worked in a simple laboratory. Candles gave a dim flickering light. The only source of heat was a small fire.

Boyle experimented with glowing wood and meat. He studied glowworms, which are not really worms at all, but the *larvae* (young feeding stage in the life cycle of most insects) of a beetle. Boyle discovered that no heat was given off from bioluminescence. He found, too, that pouring liquid on luminescence dimmed the light.

Another of Boyle's experiments had to do with oxygen, even though in the 1600's oxygen hadn't been identified. What would happen if he removed all the air from his samples, wondered Boyle. To find out, he put luminescent wood and glowworms under large glass jars. Then he pumped all the air out of the jars, creating a vacuum. He found his samples couldn't stay lit. Then, Boyle allowed air into the jars, and the wood and glowworms lit up again.

It may not sound like much, but Robert Boyle had made a major discovery. He became the first person to prove that at least one thing is necessary for bioluminescence: oxygen.

Piecing together puzzles in science can take many years. One scientist can make discoveries as Robert Boyle did. The information is shared with other scientists. They, in turn, make their own discoveries, and more and more pieces of the puzzle fall into place. Finally, someone has enough information to put all the pieces together.

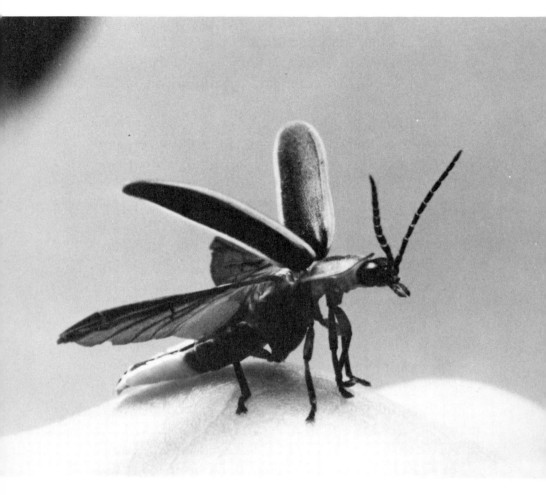
Firefly or click beetle.

It was not until 200 years after Boyle's work that the next piece of the bioluminescent puzzle was uncovered.

Raphael Dubois, a Frenchman, was interested in many areas of science. In 1885, Dubois had a laboratory by the sea, and became interested in luminous clams on the beach. He began experimenting with the luminescence of the clams, and then of click beetles.

Dubois cut out and crushed the two luminous organs of the click beetle. He mixed one with hot water and it didn't glow at all. Heat, he thought, may have destroyed something in the organ. Dubois mixed the second light organ with cold water. It glowed for a while, but soon went out. Dubois guessed that whatever had caused it to glow that short time was used up. But when he mixed the two samples, the hot and cold, they glowed again.

Dubois repeated this experiment over and over. Then he experimented with the luminous organs of the clam and got the same results.

Gradually, Dubois concluded that at least two substances besides oxygen must be present for bioluminescence. One, the substance destroyed by heat, he named *luciferase*. Dubois decided luciferase was an enzyme. Enzymes act as helpers to speed up the action of combining chemicals. They are easily destroyed by heat.

The second substance he called *luciferin*, and he concluded it was a chemical compound.

Dubois reasoned that luciferin and luciferase were both present at the start of his experiments, in both the hot and cold water. But luciferase was quickly destroyed by the heat, leaving only luciferin. And, since luciferin combined with cold water, its structure was changed and it lost its ability to glow.

Dubois also discovered something else about bioluminescence. He tried to exchange the chemicals of the click beetle for those of the clam, and it didn't work. Dubois decided that each luminescent animal has its own special form of luciferin and luciferase to make light.

At the same time that Dubois was working in France, an American, Dr. E. Newton Harvey was also hard at work on the puzzle of bioluminescence. Dr. Harvey's laboratory was at Princeton University. He used fireflies for his experiments.

Over the years, Dubois and Harvey wrote to each other, and together they added still more answers to the puzzle of bioluminescence.

They isolated luciferin and luciferase from the firefly's tail. This means that they had the pure chemicals to work with. Then Dr. Harvey and his students were able to show that these chemicals are different in nearly every luminous species, thus proving Dubois' theory.

But unsolved mysteries about bioluminescence still existed: What made luminescence bright? What turned it on and off? How did fireflies and other creatures know when to light up?

A student of Dr. Harvey added still more answers to the puzzle of bioluminescence. He was Dr. William McElroy, who worked with bioluminescence at Johns Hopkins University, Maryland.

McElroy knew how luciferin, luciferase, and oxygen worked together. He also knew about a substance called ATP—*adenosine triphosphate*, a chemical compound.

ATP controls a cell's activities. Every living thing is made up of one or more cells. So ATP controls the energy for every bit of movement and thinking you do. It controls the energy that helps plants and animals grow.

Could ATP be another chemical of bioluminescence? Could it be ATP which made luminescence bright or dim? Dr. McElroy thought it might be. So, he set up an experiment to find out.

Using scissors, Dr. McElroy and his assistants snipped the tail lights from thousands of fireflies. They used the tails to make pure luciferin, luciferase, ATP, and another chemical, magnesium sulfate.

Then they tested different mixtures of the four ingredients. Tests were made in the open air so there was a constant supply of oxygen.

In test after test, when enough ATP was added, the mixture always gave good light. ATP was proved responsible for the brightness of bioluminescence.

And Dr. McElroy became the first scientist to make bioluminescense in a laboratory.

The world of bioluminescence has grown since McElroy's experiments with ATP. More plants and animals which luminesce have been identified and different chemical systems which produce light have been discovered.

Dr. McElroy and his assistants used scissors to snip the tail light off of thousands of fireflies.

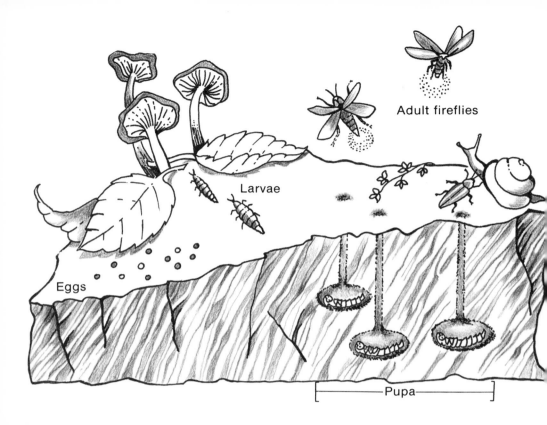

After mating, the female firefly deposits her eggs on the ground. In a few days, tiny worm-like larvae hatch from the eggs. They feed on tiny organisms in the soil, then on bigger and bigger pieces of food until they eat animals bigger than themselves. Snails, slugs, and earthworms are often on the menu. The larva injects a kind of poison into its prey changing the victim's body tissue into a soft, almost liquid state for easy eating. After much feasting, the firefly doesn't eat again, but lives on protein stored in its tissues. When the weather turns cool, the larva burrows into the ground and gets a new look and name: pupa. When it's warm enough, the pupa emerges from the ground as a full-grown firefly — and begins flashing for a mate.

# CHAPTER 3

# Flashing Firefly

Fireflies live almost everyplace on earth where there is at least one warm season a year. In the United States, fireflies are sometimes called lightning bugs, or glowflies. Trinidad fireflies are called candleflies. In Puerto Rico, they're cucobanos. Small fireflies in Jamaica are called blinkies, and the larger ones are known as peeneewallies, or click beetles.

Women who live in the Caribbean area sometimes pin click beetles to their clothing or hair. Or, they might hang one on a chain for a necklace. In the tropical rain forests of Central and

South America, people sometimes attach click beetles to their shoes to help light wooded paths for them at night.

In Japan, special days were set aside to honor the firefly. Today, air pollution has caused most Japanese fireflies to disappear. But, fireflies are still important to the Japanese. Hundreds of the small insects are raised in the country by the Japanese government. To mark the start of summer, the fireflies are taken to Tokyo. At a special night ceremony, the fireflies are released as children and adults alike shout and clap. The firefly is a new symbol in Japan. It stands for clean air.

The firefly is an insect, one division of the animal kingdom. The insect division is made up of many groups. Fireflies belong to the beetle group.

All beetles have two sets of wings, and their bodies have three parts—the head, thorax or chest area, and abdomen. And, they have antennae, or feelers.

The beetle group is further divided into families, or species. Fireflies belong to two species. One is called *Elateride*, and the fire or click beetles living in the hot forests of Central or South America belong to the *Elateride* family. The most common firefly belongs to the *Lampyride* family. The *Lampyride* is much smaller than the *Elateride* and lives in North America.

All beetles grow in a similar way. And you might say a firefly's life begins with a sparkle in some other firefly's tail. Experts agree that the firefly flash has but one purpose—to attract a mate. Mostly, the flashes come from the females. Sometimes the male flashes for a mate, and the female answers with a flash a few seconds later.

Each species of firefly has its own pattern of flashes. There are differences in the color, strength, and length of each flash. It is a kind of code. Fireflies of different species usually do not respond to each other's flashes. And, sometimes a female will reject a male even if he is of the same species.

If a male is interested in a female, he follows her flash. More than one male may answer the female and when this happens, the males fight until one is either killed or driven away.

The adult firefly has only from twenty-four hours to about two weeks to live. How long depends on the type of firefly, the weather, and other conditions where it lives.

Because of its sparkle, the flashing firefly

could be easy prey for birds and other insect-eating animals. But when the firefly senses danger, it covers its body with a terrible tasting liquid. At the first taste of this "blood," the animal usually spits the firefly to freedom.

One firefly, the female *Photinus pyralis*, is an enemy to male fireflies. Once pregnant, she no longer responds to males of her own species. However, the *Photinus pyralis* can mimic females of at least three other species. She uses this ability to become a killer. When pregnant, she lures males of other species with the flashes, and then pounces on them and kills them.

Scientists have not yet figured out why such aggressive behavior exists. But *entomologists*, scientists who study insects, are looking into the puzzle. They believe the tiny firefly may have a far more complex brain than was once believed possible.

The common North American firefly is found from Canada to Florida, and from Kansas to the East Coast. No fireflies live west of the Rocky Mountains. If you want to take a closer look at fireflies, the best time to find them is from late May to August. Fireflies are the most active just at sunset. Then go hunting in a yard, field, or near bushes.

Because the firefly's light is a chemical reaction, the outdoor temperature will affect its frequency. The cooler the air, the longer each flash will be and the time between flashes will be longer, too. When it gets hotter, flashes will be shorter and occur more often.

Fireflies continue to be an important part of scientific research and the work medical doctors, biologists, and entomologists do.

# CHAPTER 4

# Sea Fire and Other Water Lights

Have you ever seen sparkling light in the ocean at night? If so, you've seen dinoflagellates. They live in all of earth's oceans.

*Dinoflagellates* are one-celled organisms that live in water. They are so tiny they can only be seen through a microscope, unless, of course, they are lit up. Dinoflagellates are very sensitive to the movement of water, and scientists think they may light up to show alarm. They flash a bright blue light, a new flash appearing every one-tenth of a second. Scientists do know that dinoflagellates use a luciferin and oxygen chemical system to make their light.

Every once in a while, dinoflagellates have a population explosion. Millions and millions of new-born organisms turn the ocean a brownish-red color. This is called a *red tide*. Red tides occur regularly in both the Atlantic and Pacific Oceans. The Red Sea is named for them.

Perhaps you can imagine the beautiful sight of a red tide at night, the sea aglow with bright blue lights. In ancient times sailors called the sight *sea fire*. They didn't realize that the lights were cold, without the heat of firelight.

Though beautiful, red tides can be dangerous. Some species of dinoflagellates make a strong poison. The poison kills fish and is passed to the birds that eat the poisoned fish.

*Cypridina*, sometimes called a *sea firefly*, is also luminescent. *Cypridina* is a crustacean, which means it has an outer skeleton — its shell. It is about the size of a tomato seed, and *Cypridina* is so light in weight that it takes one-half million of them to make one pound. If it is crushed, and water is added to it, *Cypridina* give off a bluish glow. Scientists like to work with *Cypridina*. *Cypridina* luminescence lasts for a long time. Even 25 years after dying, *Cypridina*

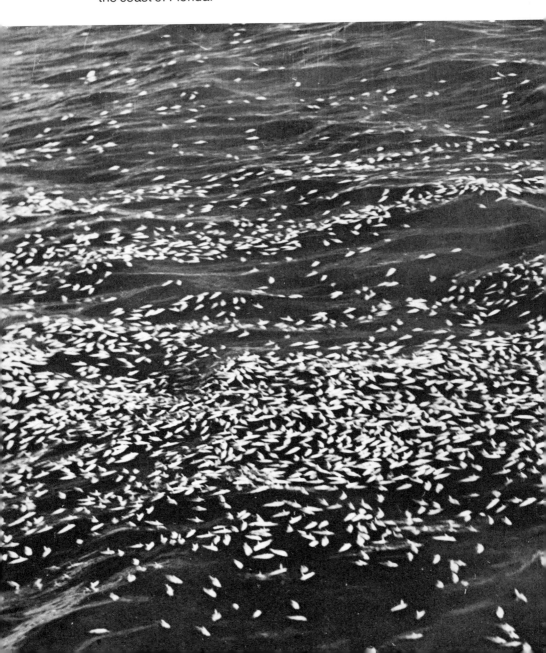

Sparkling red tides may kill many fish, such as these off the coast of Florida.

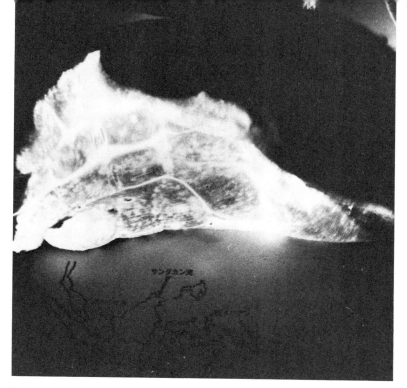

Dr. Johnson has spread *Cypridina* on his hand, moistened it with water, and can read the map in the dark.

can be used for study.

*Cypridina* gives off enough light to read by. Japanese soldiers used *Cypridina* this way in World War II while fighting on the islands of the South Pacific Ocean. Hiding from their American enemies, they needed to read their maps at night. The soldiers crushed *Cypridina* in their hands, spit on it, and used the resulting luminescence for light.

Dr. Frank Johnson is seen in the light from a beaker of *Cypridina*. To get the light, he added water to the beaker, which contained thousands of the animals.

Even before World War II, there was interest in *Cypridina*. Dr. E. Newton Harvey, whose work with fireflies you have read about, also worked with *Cypridina*. In fact, they were his favorite subjects. He first came across them on a trip to Japan in 1916. *Cypridina* live in the sand of shallow water along the Japanese coast. Dr. Harvey had many pounds of *Cypridina* sent to Princeton University.

Dr. Harvey found that *Cypridina* makes its luminescence in two separate glands. One gland makes luciferin, the other luciferase. When *Cypridina* "spits" these chemicals out, or when they are squeezed out by crushing, they combine with water to make light.

A student of Dr. Harvey's, Dr. Frank Johnson, went on to teach at Princeton and to contribute to the body of scientific knowledge about luminescence. Dr. Johnson is most proud of his work with a beautiful jelly fish, *Aequorea*.

For many years, scientists believed the luciferin-luciferase chemical system was the *only* chemical system which could produce light. Dr. Johnson discovered that more than one type of

The beautiful jellyfish *Aequorea* helped scientists discover new chemical systems that make luminescence.

chemical mixture can make luminescence. He found this new mixture in the *Aequorea.*

*Aequorea* live in Pacific waters off the coast of Washington and British Columbia. They grow to between three and six inches in diameter. They have a strip of luminescent rings which glow at the outer edges of their round bodies.

*Aequorea* are so numerous, that they become tangled in fishermen's nets, and were considered little but a nuisance. That is, until Dr. Johnson came along.

Dr. Johnson spent many summers at an outdoor laboratory at Friday Harbor, Washington. With the help of local teenagers, he collected tons of *Aequorea*. Dr. Johnson and his assistants had to remove its luminous "ring." At first this was done by cutting it off with a scissor. Later a machine was developed which made the job much easier.

After a summer of collecting *Aequorea* samples, Dr. Johnson sent the luminous rings to Princeton. Then he went to work to answer the question, "What makes the jelly fish light *its* light?"

Dr. Johnson found that *Aequorea* have a special type of protein within their bodies called a *photoprotein*. A photoprotein can give off light under certain conditions. Dr. Johnson went on to find out what these certain conditions were in *Aequorea*.

He named the photoprotein *aequorin*. Dr. Johnson discovered aequorin will produce light in combination with calcium. Aequorin also lights up with the radioactive element, strontium. In the jelly fish, tiny amounts of calcium are released, combine with aequorin, and light results.

As a result of Dr. Johnson's work, two more creatures were discovered which light up because of combinations of aequorin and calcium. One is the sea pansy, *Renilla*, a pinkish animal about as big as your thumb. The other is the hydroid, *Obelia*. Hydroids are animals that look like plants. Sea pansies, hydroids, and jelly fish are all related.

It may look like a plant, but this hydroid is really an animal related to corals and jellyfish.

The luminous squid has sparkling blue-white lights all over its body and two lantern-like eyes which only light up in the dark. The squid's and other sea creature's lights are sometimes called *photophores*.

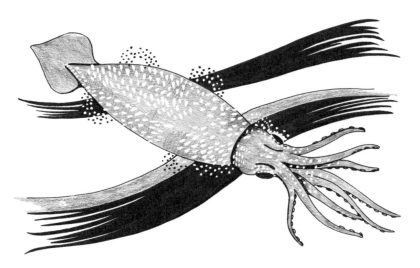

This luminous Japanese squid is only the length of your finger, but it makes good eating. Every spring, on the west coast of Japan, thousands of these squid come to the surface to mate. When it's dark, they begin shooting through the water at great speeds. Crowds of men, women and children in boats scoop them up in nets, and later have a feast.

So far, you've read about animals which luminesce in shallow sea, or salt water. Scientists know of only one freshwater animal which luminesces. It is a limpet, *Latia*. All limpets are mollusks, animals without backbones. Snails, clams, and limpets are mollusks, and live inside a shell.

*Latia* are found in the cold, rushing waters of New Zealand's mountain streams. They cling to the bottom of rocks with a kind of suction cup. Scientists collect the limpets, and then keep them frozen until they return to their laboratories.

To luminesce, *Latia* secrete a bright green slime, using the luciferin-luciferase system to make their light. Little is known about the reasons for *Latia*'s luminescence. Little is known about why only one of many freshwater animals luminesces. But, scientists are working on the puzzle.

LATIA

Scientists wade in the cold waters of New Zealand's streams to collect *Latia*.

# CHAPTER 5

# Deep Sea Lights

It was a beautiful day in 1930. Marine biologist William Beebe and engineer Otis Barton stepped into a hollow steel ball which was just big enough for the two of them. The door of the ball, a *bathysphere*, was locked shut, and the men looked out of the large quartz window in its side. When they gave the signal, the bathysphere was carefully lowered over the side of their ship, anchored off the Bermuda coast.

It was a test run of this new diving equipment. Otis Barton had designed the bathyspere to withstand great pressure from the water surrounding it. William Beebe was there to study life in the ocean.

William Beebe in the bathysphere.

The scientists knew they could telephone the ship if trouble should occur. But still they wondered. Would the great water pressure in the ocean depths crush the bathysphere, and the men inside it?

They descended . . . 1,000 feet, 2,000 feet . . . no one had ever been this deep in the ocean before. Still they went deeper, and finally stopped at a little more than 2,500 feet.

The scientists were able to see out of the bathysphere with the help of large searchlights attached to it. William Beebe and Otis Barton made history that day by being the first persons to go down one-half mile into the ocean's depths. Not only was the bathysphere a success, but Beebe opened up a whole new world of bioluminescence to scientists. He reported seeing many bioluminescent fish no one had known of.

Today, scientists have identified more than one hundred species of deep-sea fish which luminesce. Let's take a look at some of them.

Among the smallest are the hatchet fish, so named because they are shaped like a hatchet blade. Hatchet fish are only from one-quarter inch

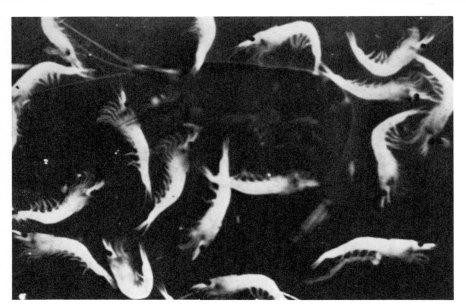

Krill.

These bioluminescent shrimp and the krill swim deep in Earth's oceans.

to two inches in length. They light up on the underside of their bodies.

Another small bioluminescent fish is the lantern fish. This fish not only has lights along the side of its body, but it also has two bright lights, like lanterns, on its tail.

Scientists know these little fish make their own chemicals to light their lights, but they do not know what chemical systems are used as yet.

Other fish about which little is known are the luminescent shark and the spiny dogfish, which is a type of shark.

Scientists *do* know a great deal about a large group of deep-sea fish which have light, but don't make the light themselves. Instead, light is produced by a luminous bacteria which grows on them.

You may remember, bacteria are tiny plants which can be seen only with a microscope. These bacteria live by feeding on other plants, or animals. Such a relationship is called *symbiotic*. This means it is a relationship that is of benefit to both the bacteria and the fish. Bacteria get food, and fish get lights.

Biologist J. Woodland Hastings and a team of twenty other scientists from the University of

California recently traveled to the Banda Islands in the South Pacific Ocean to study the pony and flashlight fishes, which both show bacterial luminescence. Their ship, *Alpha Helix*, had a laboratory on board, complete with fish tanks and many instruments, including one which measures the strength of different kinds of light—*a photometer*. The scientific team caught and kept pony and flashlight fish in the tanks so they could be studied.

*Flashlight fish* are just a few inches long and have two lights that look like flashlights, one under each eye. Bacteria live in sacs under the eyes. When the fish needs to "flashlight," it simply opens the eyelid-like covers over these sacs.

*Pony fish* have snouts which are shaped like a pony's nose. Light comes from bacteria which live on its underside.

A blue light on the ceiling of *Alpha's* laboratory could be brightened or dimmed. It made the inside of the laboratory look like daylight or night time under water.

The photometer measured the strength of the overhead light. Then, it measured the effect of this light on the fish.

It was found that the brighter the overhead

light, the brighter the flashlight's luminescence. Dr. Hasting's team concluded that the ability of the flashlight fish to match its own light to surrounding light was camouflage. It could help the fish hide from predators swimming below it.

Dr. Hasting's team was not the only one interested in the flashlight fish. At about the same time, another team was at work in the warm waters of Israel's Gulf of Elath. James G. Morin and his wife, Anne Harrington, were on the team. They are biologists and underwater photographers. Part of their job was to get photos of flashlight fish behavior under water for the University of Southern California.

Morin and Harrington were surprised to find they did not need to use their underwater lighting equipment to get good photos of the fish. Swimming in groups, the flashlight's light was strong enough.

Another deep sea fish which uses bacteria for light is the angler fish. Fishermen are sometimes called *anglers*, and the angler fish is a good imitator of a fisherman. Coming from the top of its head is a long rod-like organ. At the end of this "rod" is a luminous bulb called an *esca*. Luminous bacteria grow in the esca, and the light attracts

This strange looking fish is the female angler. At the end of her "fishing rod" is the esca.

other fish. If they come to investigate the blue-green luminescence, they quickly become a meal for this fisher fish.

Only the female has a luminescent lure. Male anglers are only about one-tenth the size of the female, and along with bacteria have a symbiotic relationship with the female. After birth, the male swims about until he finds a mate. Then he attaches himself to her by boring anywhere into her flesh. The male angler is a parasite, living off the female. He depends so much on her, that he eventually loses most of his body organs. Finally, he can do nothing but fertilize his mate's eggs when they are laid.

Still other creatures have "fishing rods" they use to lure their prey. Two fierce-looking examples are the dragon and viper fish. These fish have "rods" with bulbous escae which light up, and the sides of their bodies also luminesce with hundreds of lights.

The scarlet red shrimp have several escae protruding from their "fishing rods," as well as hook-like spines. The escae lure prey, then the spines catch it and transfer the food to similar hooks on the shrimp's legs. The legs then bring the food to the shrimp's mouth.

# CHAPTER 6

# Plant Luminescence

Bacteria are plants, and you have read how luminescent bacteria can live on various fish. Luminous bacteria can also live on other animals and plants as well.

Bioluminescence caused by bacteria is usually found near water. And sometimes—like Robert Boyle's chicken—it might be where dead fish or meat are not refrigerated. Before antibiotics were developed, bacterial luminescence sometimes grew on people, in open wounds.

During the American Civil War, in the 1860's, doctors reported seeing soldiers' wounds aglow. Imagine the eerie sight created by glowing bodies at night. It is easy to understand how stories of battlefield ghosts might have circulated.

Doctors in those days thought it was a good sign if a wound glowed. They believed it meant wounded body tissue was being eaten by the bacteria. Luminous bacteria are not harmful to humans, and luminescent wounds seemed to heal better and faster.

Bacterial light is given off in a steady, yellow-green or blue-green glow. It takes millions of bacteria to make even a small amount of luminescence. Bacteria are the smallest of luminous organisms, and some species are only 1/20,000 of an inch in diameter.

Scientists are in the process of working to isolate the chemicals which go into bacterial luminescence. They know more than one chemical is involved, and that an enzyme like luciferase helps create the light.

Besides bacteria, another large group of plants luminesce. These are the *fungi* (singular: *fungus*). They have no flowers, stems or leaves, and do not use the sun's energy to make their food as all other plants do. Most are parasites on other plants or animals. Mushrooms and toadstools are examples of fungi. Scientists have identified several dozen species of luminous fungi. They do not yet know the chemistry of fungi luminescence.

Jack-my-lantern toadstools by day, above, and by night, below.

Since fungi can't make their own food, they get it from other sources, dead or alive. Fungi do this by inserting thread-like roots called *mycelia* into the food. If the food is dead wood, and it often is, the wood might glow with a dull green or yellow light. This glow is sometimes called foxfire, and has been considered magical by some people. Natives of some South Pacific islands have been known to smear their faces with foxfire to frighten their enemies in battle.

Mushrooms and toadstools are fungi which are often found growing on decaying plant matter. One luminous species, *Mycena*, gives off a glowing green light which comes from gills under its cap. *Mycenae*, are also known as Jack-my-lanterns.

A mushroom, *Lampteromyces japonicus*, or Moonlight mushroom, looks like a flower by daylight, and even stranger when lit up.

One fungus uses a living plant for food. It's the *Armillaria* and it lives on coffee trees in South America. *Armillaria* has been known to light up a whole coffee plantation at night. It may be a pretty sight, but *Armillaria* actually does great damage to the coffee trees.

The moonlight mushroom is a tan color during the day, left, and glows a bluish-green color at night, right.

# CHAPTER 7

# Why Light Up?

Why does bioluminescence occur? Attracting prey is one reason.

In New Zealand's Waitomo Cave, thousands and thousands of glowworms hang from the cave's ceiling like stars in the sky. Glowworms aren't really worms at all. They are the larvae of two kinds of beetles. One of these is the firefly. The larvae spin a sticky material into thread-like lines, perhaps as many as 70 per "worm." They use these lines to trap their food. Insects are attracted to the shiny lines, investigate, and get caught in the threads. Then the glowworms reel in their dinner by drawing up the lines. When the insects are close enough, the glowworms gobble them up.

Hundreds of visitors come each year to see the glow-worm cave in Waitomo, New Zealand.

A closer view of the glowing worm as it hangs waiting for an insect to become caught in the shiny lines it has spun.

The dragon fish has long, sharp teeth which it uses to catch prey attracted to the two rows of lights along its side.

Some fish also use their lights to attract prey. Flashlight fish and angler fish are examples. Swimming in schools, they make pools of light which attract smaller fish, their main source of food. Again, when the prey is within grabbing distance, it is gobbled up.

Flashlight fish light up for another reason, too. They flash their lights to defend themselves. Swimming alone, a flashlight fish might see an enemy coming dangerously close. The flashlight opens its lightlid, and the enemy is attracted to the light. Then, the clever flashlight uses a trick. It closes its lightlid, and darts away in the safety of darkness.

The flashlight fish may also defend themselves by flashing their lights on and off as they

Here the flashlight fish has its light lid open.

Angler fish come in different shapes and sizes. This one looks fierce, indeed.

swim in a zig-zag pattern. This makes them hard to locate.

*Cypridina* also uses its light to defend itself. When it is threatened, *Cypridina* "spits" luciferin and luciferase into the water and light results. The light is in the water and not on the *Cypridina*'s body, which is a big advantage. Its predators are attracted to the light, giving the *Cypridina* a chance to quickly swim away.

Certain deep-sea squid and shrimp have a

similar ability. Near the ocean's surface some squid make a black "ink" and use it to hide themselves. In the ocean's depths, where it is very dark, the ink is luminescent. One squirt allows the squid or shrimp to escape would-be enemies. Like the *Cypridina*, they simply squirt and run.

Bioluminescence is used defensively in still another way. Some fish can control the amount of luminescence they give off. Remember the pony fish? It matches the amount of light coming from the underside of its body to the amount of light in the water. The pony fish has a covering over its bacterial luminescence which it can adjust to give the amount of light it needs to blend into the surrounding water better.

Communication is another reason for luminescence. Courtship and mating signals are one form of communication. This communication becomes spectacular in Thailand and Malaysia, two countries in Southeast Asia. There, male fireflies swarm together in large trees. When the sun sets, they begin to flash . . . first one, then another and then another. Soon a whole tree is flickering.

Then, the males all begin to flash at the same time. Lights go on and off at exactly the same second.

Scientists aren't certain why these fireflies flash in this way, but they think it may be to give the females a very definite idea as to where the males are.

Another brilliant display of communication takes place near Bermuda and in the Caribbean islands. It happens only in summer, exactly two nights after a full moon, and about 40 minutes after sunset. In the shallow water over coral reefs, the fireworms come to the surface to mate. Fireworms are a species of annelid worms, the most highly developed of worms.

Females gather on the water's surface and begin to swim in small circles. They leave a trail of blue or yellow-green slime. At the female's light signal, males, too, come to the surface. The males give off short flashes of light as they join the females. Many swimming groups form, and the circles become smaller and smaller as the worms swim closer together. Females drop eggs and the males fertilize them with sperm. When the mating is complete, the worms dive back into deeper water.

Christopher Columbus was probably the first person to write about fireworms. He didn't know what they were, but in his ship's log he wrote about seeing candles on shore. Experts agree the candles were probably fireworms.

One of the most beautiful creatures to luminesce for communication is the female railroad worm. Like the glowworm, it's not really a worm at all but a wingless beetle, *Phrixothrix*. Railroad worms are related to fireflies. Found in Central and South America, the railroad worm looks something like a lighted train. It has a red headlight, and eleven pairs of green lights glowing along its side.

# CHAPTER 8

# Cold Light and Medicine

A doctor and nurse watch Sally Braun on the hospital bed. The 13-year-old's face is flushed and she feels terrible. Sally has a serious bacterial infection which has made her temperature rise to 106° F.

"There's no question about it," the doctor says. "The antibiotic we gave Sally isn't working. We'll have to try something else."

The nurse frowns and says, "Yes, but it will take at least 24 to 48 hours to find out if another

antibiotic will work. By that time, Sally could suffer serious damage from this fever."

"Not if we use the firefly test," the doctor answers. "We'll get a blood sample and tell the lab to be ready. In just two hours we can find out if a different antibiotic will destroy this infection."

Sally's blood sample is sent to the laboratory. It is mixed with chemicals from fireflies' tails. Results from the test help technicians decide which antibiotic will kill the harmful bacteria. They also find out how much medicine it will take to cure Sally. The doctor gets test results quickly, and shortly after taking a new antibiotic, Sally's temperature begins to lower. For her, watching flashing fireflies on a summer's evening will probably always have a special meaning. Fireflies helped make her healthy again.

The story of Sally Braun is imaginary, but it could be true. Firefly tests, are very real and are being used more and more across the United States. The test was developed by biochemist Dr. Grace Picciolo, and a team of scientists at the Goddard Space Flight Center, Greenbelt, Maryland.

Dr. Picciolo knew fireflies have the only known bioluminescent system which needs adenosine triphosphate to make light. Every living

Dr. Grace Picciolo helped develop the Firefly Test. Here she works with a vial of luciferin and luciferase.

thing, from bacteria to human beings, makes ATP. Where ATP exists, some form of life is sure to exist as well.

Dr. Picciolo also knew that unhealthy cells produce less ATP than healthy ones. And, sometimes they produce more ATP. If body cells are cancerous, for instance, they make *less* ATP. Harmful bacteria, on the other hand, produces *more* ATP. Knowing how much ATP is being produced can help doctors diagnose disease, as well as decide the amount and type of medicine which needs to be prescribed.

A sample of patient's blood is taken. Chemicals are used to separate the bacterial ATP from the blood. The ATP is mixed with fireflies' luciferin and luciferase. Light results. The important question at this point is, "How *much* light is produced?"

To find out, the technician puts the mixture into test tubes which are attached to a photometer. The photometer measures the strength of the light. The more ATP in the blood, the more harmful bacteria are in it, and the brighter the glow will be.

Once the amount of bacteria in the cells has been determined, the effect of different antibiotics on the growing bacteria can be measured. Antibi-

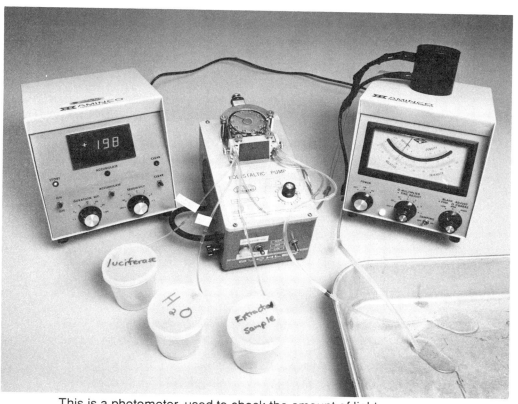

This is a photometer, used to check the amount of light given off by firefly chemicals and the ATP in human blood.

otics are added to bacterial ATP-luciferin-luciferase mixtures. After they've had a chance to begin destroying the bacteria, the test tubes are again attached to the photometer. The amount of light tells the technician how well each antibiotic is doing its job. And the doctor learns which medicine is the best medicine.

Fireflies also help diagnose heart attacks. One specialist in such diagnosis is Dr. Marlene De Luca. She works at the University of California, in San Diego. As part of her work there, Dr. DeLuca has developed a quick heart attack test using firefly chemicals.

All vertebrates, animals with backbones, including human beings, have small amounts of a chemical called CPK (*creatine phospho-kinase*) in their muscles. CPK produces ATP.

When muscle cells become unhealthy, large amounts of CPK are released. This also means large amounts of ATP are produced.

The heart is a muscle, and some heart attacks are caused by unhealthy heart muscle cells. To test for an attack, a small amount of a patient's blood is taken. The blood is mixed with fireflies' luciferin and luciferase. The worse the heart attack, the more CPK is in the blood. The more CPK, the more ATP . . . and the brighter the light of the mixture.

Unhealthy muscle cells in other parts of the body cause other diseases as well. Muscular dystrophy is one. Dr. DeLuca's test makes it possible to test for muscular dystrophy in infants. It is a painless test, as only a drop of blood is needed.

Early detection of muscle disorders is important as it gives doctors a better chance to help patients. It also gives patients a better chance to help themselves.

Aequorin may also have many uses in medicine in the future. Since aequorin lights up when combined with calcium, a lack of calcium in the body could be detected in a test similar to the firefly test. Calcium is an important ingredient to normal functioning of human nerves, muscles, bones, and other tissues.

# CHAPTER 9

# Cold Light and Technology

Along with their uses in medicine, firefly tails also help other scientists in their work. They've even been to Mars!

Scientists at the Goddard Space Flight Center, including Dr. Grace Picciolo and Dr. Emmett Chapell, developed a life detection system called "Project Firefly." Project Firefly was to go aboard the Viking 1 spaceship which landed on Mars on July 20, 1976 to help answer the question, "Is there life on Mars?"

Scientists reasoned that if ATP was present in all forms of life on Earth, there was a good chance this might be true in space as well. So, they

As part of the Goddard Space Flight Center's "Project Firefly," Dr. Emmett Chappell uses the most modern scientific equipment in his work.

programmed an instrument to scoop up a small amount of Martian soil, grind it into dust, and add firefly chemicals ... *except* for ATP. If light resulted, it meant ATP was on Mars. If ATP was on Mars, then life was on Mars. A photometer had been developed to measure the light from one trillionth (1,000,000,000,000) of a gram of ATP. Another instrument would change the light energy to electrical energy, and results of the light detection test would be sent to tracking stations on Earth.

As it turned out, Project Firefly was not used on Mars. Another life detection system was used and found no evidence of life.

But, other uses have been found for "Project Firefly." Besides its use in medicine, the firefly test now helps citizens check for water pollution in their lakes, rivers, and reservoirs. A great deal of water pollution is caused by harmful bacteria. Harmful bacteria in water behaves as it does in blood, and produces a lot of ATP. If the light from the firefly test is too bright, scientists know that the water is unsafe for drinking, or swimming.

So far, the use of aequorin has been limited because of its short supply. Few people besides Dr. Frank Johnson have spent the time necessary to collect the jellyfish. But more and more scientists are capturing jellyfish for their own supplies of aequorin. It is expected that aequorin will be used to check such things as the amount of calcium present in milk.

Aequorin also lights up in the presence of the rare metal strontium. Strontium is present in radioactive fallout. Therefore, aequorin could also be used to check for fallout in nuclear generating plants, or in the event of a nuclear bomb explosion.

Luminescent Aequorea.

# CHAPTER 10

# Helping Science

Each year many thousands of fireflies give up their tails to science. Hundreds of young people earn money by catching them. Firefly hunters also enjoy knowing they help with scientific research.

Bees or ants can be farmed, but fireflies must be caught. Those bred in captivity don't produce enough of the needed chemicals. Scientists have learned how to make an artificial form of luciferin. But the firefly is the only source of the enzyme luciferase.

Perhaps you would like to get in on the action and become a firefly hunter. Here's what to do:

- First contact a company which will buy your firefly catch. Today, two United States companies buy fireflies. Each company has special rules for the care of the insects, and you should know what the rules are *before* you catch the insects.

>Antonik Laboratories
>P.O. Box 15
>Elk Grove, Illinois 60007

Antonik buys fireflies *only* if they can be picked up by truck. This means you need to live near Elk Grove.

>Firefly Club
>Sigma Chemical Company
>P.O. Box 14508
>St. Louis, Missouri 63178

Sigma Chemical Company buys fireflies from any place in the U.S. They supply special cans for mailing the insects to Missouri.

BE SURE TO WRITE TO THESE COMPANIES FOR MORE INFORMATION BEFORE MAILING FIREFLIES TO THEM.

• From May through November is when you can find fireflies in North America. The exact months vary with the climate. They are easiest to catch at sunset, and they are easiest to catch with a net. Using a net also prevents injury to fireflies.

If you don't have one, making a net is easy!

1) Make a hoop on one end of a piece of wire. You can straighten out a wire hanger for this. You will attach this piece to the handle.

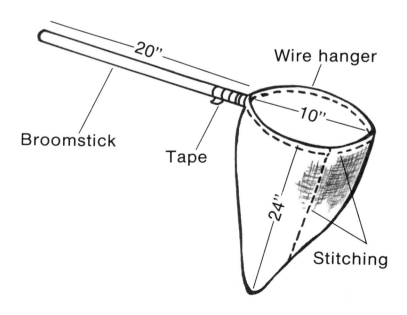

2) Get a piece of cloth or netting to fit the hoop. Sew it into a bell bottom shape. The wide bottom helps you shake the fireflies into it.

3) Sew the cloth onto the hoop.

4) Tape the straight piece of wire to a stick or broom handle. The handle should be about 20 inches long. Be sure all the wire is covered with tape to prevent injury to your hands.

- Do not squash the insects, as they are bought only if they are in good condition. Don't use any chemicals with them. Adult fireflies don't eat, so it isn't necessary to give them grass or leaves.
- Be sure to keep track of how many fireflies you catch. The easiest way to do this is to count as you catch. Then you won't have to handle the fireflies to count them later.
- You may find several species of fireflies on your hunt. You should know that scientists like best to use a male called the North American Black. Its chemicals are the strongest. The North American Black has outer wings which are black with a red rim. Its light is a bright lemon-yellow color.

You may also find a small pale firefly and the North American Brown firefly which has wings of a brownish color and a pointed dull yellow light. If you can, avoid the Brown and pale fireflies as they are not as effective in research work as the North American Black.

- It might be tricky to do, but if you can avoid catching female fireflies, you should. The females need to be kept alive so that they can lay the eggs of future fireflies. You can tell a female because it is larger than the male and has a smaller light.
- When you've caught a net of insects you will want to transfer them into a container, like a quart jar that has a lid. A simple trick will help you do this. First shake the insects to the bottom of your net, and fold the net in half so they can't escape. Then put the net into the refrigerator. Cold fireflies become sleepy and are easy to control.
- If you do want to put active fireflies into a jar, here's another trick you might use:

Use a wide-mouth jar, and make a funnel for it out of a piece of paper. Cut a ½-inch hole in the end of the funnel. Tape the side of the funnel and put it into the jar. Cut away any excess paper on the top, and tape the funnel to the jar.

Tape

Clip off to make ⅜" hole

Push the jar upside down into the net. To manage this, support the net over two chairs.

Then, watch the fireflies crawl up the funnel. Fireflies always crawl up, never down. Once they leave the net, they can't get back into it.

Leave the funnel in the jar. You can use it again the next time you go firefly hunting. Put the lid on the jar and be sure to keep your catch in the refrigerator.

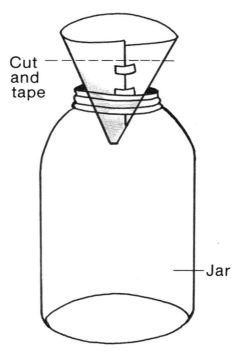

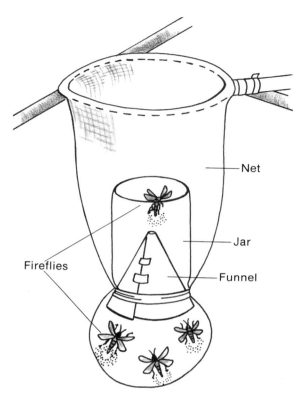

• When you have caught the required number of fireflies, the next step is to get your collection to the buyer. It is important for you to use the information the buyer has sent you at this time to insure safe delivery of your catch.

• You will be paid for the fireflies depending on how many you are able to deliver. But your biggest reward will probably be knowing that you are helping science. Go to it, and happy hunting!

# The Bioluminescent Plants and Animals of LIVING LIGHT

| Type of Organism and Common Name | Color of Light | Description and Reason for Luminescence | Chemistry |
|---|---|---|---|
| Annelid<br>Fireworm | blue or yellow-green | Leave a trail of slime. Reason uncertain, may frighten predators. | Uncertain |
| Bacteria | blue-green | Steady bright glow. Reason not known. | Uncertain |
| Coelenterates<br>Aequorea-jellyfish<br>Hydroid<br>Sea Pansy | greenish-blue | Rapid flashes. Reason is to frighten predators. | Photoproteins stimulated by calcium. |
| Crustaceans<br>Cypridina<br>Shrimp | blue | Squirts chemicals into water. Diverts or frightens predators. | Squirt luciferin, luciferase into water. |
| Dinoflagellates | blue | Rapid flashes. Reason uncertain, perhaps to frighten predators. | Luciferin and oxygen |
| Fungi, Mushrooms<br>Jack-my-lantern<br>Moonlight mushroom<br>Coffee tree fungus | yellow-green | Steady dim glow. Reason unknown. | Unknown |
| Fish<br>Dogfish shark<br>Luminous shark<br>Pony fish | green | Light up at water's surface. Reason unknown | Unknown |
| Flashlight fish<br>Angler fish<br>Dragon fish<br>Viper fish | blue-green | Control of light with "lightlids." Reasons vary. | Bacterial |
| Hatchet fish<br>Lantern fish | blue | Rapid flashing | Unknown |
| Insects<br>Firefly | yellow-white | Rapid flashing. Reason is communication. | Luciferin, luciferase, ATP |
| Railroad worm | red and green | Glows. Reason is communication. | |
| Glowworm | blue | Reason is to attract prey. | |
| Mollusks<br>Limpet<br>Squid<br>Clam | green<br>green-red<br>blue | Emit slime. Reason unknown.<br>Emit "ink." Reason unknown.<br>Light comes from photopores. | Several different types, chemistry uncertain. |

# GLOSSARY

**ATP (adenosine triphosphate)**—a chemical compound which controls the action of living cells. It is found in all living matter.

**Annelid**—the most highly developed species of worm.

**Bacteria**—microscopic plants. They are the smallest of all plants, and usually consist of only one cell.

**Bathysphere**—a submersible steel vessel in the shape of a ball, used to explore deep-sea life.

**Biologist**—a scientist who studies living things.

**Bioluminescence**—light made by living plants and animals.

**Coelenterate**—an invertebrate animal, one without a backbone, such as jellyfish and hydroids.

**CPK (creatine phosphokinase)**—a chemical compound, found in humans, which produces ATP.

**Crustacean**—an animal of fresh or salt water, with a shell-like skeleton on the outside of its body, similar to an insect's.

**Dinoflagellate**—a one-celled shelled organism found in the ocean.

**Entomologist**—a scientist who studies insects.

**Enzyme**—a substance produced by living cells which activates or speeds up chemical action of the cells.

**Esca (plural: escae)**—a bulbous extension on some fish, containing luminous bacteria which acts as a lure.

**Fungus (pl. fungi)**—a plant that produces neither flowers, stems, or leaves, and does not use the sun's energy to make its food.

**Invertebrate**—animal without a backbone.

**Larva (plural: larvae)**—the wingless feeding form of young insects after they have hatched from eggs.

**Mollusk**—an invertebrate which lives in a shell.

**Mycelia (plural: mycelae)**—a thread-like root which fungi insert into their source of food.

**Parasite**—an organism that lives on or in another organism, and depends upon it for its basic needs, such as food. Some parasites are harmful to the host—the one being lived on or in.

**Photometer**—an instrument that measures the strength of light.

**Photophore**—the organ in certain fish and squid from which light comes.

**Pupa (plural: pupae)**—the insect stage during which the insect burrows underground and grows into an adult.

**Red tide**—a population explosion of dinoflagellates which turn the water brownish red.

**Symbiosis**—a relationship between the organisms that is beneficial to both.

# INDEX

## A

adenosine triphosphate, 23-24, 69-71, 75-76
Aequorea, 38, 76-77
Alpha Helix, 49
angler fish, 50-51
annelid worm, 65
antibiotics, 69-70
Armilleria, 56

## B

bacteria, 15-16, 53-54, 67-69
Barton, Otis, 43
bathysphere, 43
Beebe, William, 43
beetles, 28-29
bioluminescence, definition, 13, 15-16
blinkies, 27
Boyle, Robert, 13, 18-19

## C

calcium, 39, 73, 76-77
Caribbean, 27, 65
clam, 16, 21-22
click beetle, 21-22, 27
coffee trees, 56
Columbus, Christopher, 66
communication, 63-66
creatine phospho-kinase, 72
cucobanos, 27
Cypridina, 33-36, 62

## D

defensive behavior, 61-62
DeLuca, Dr. Marlene, 72
dinoflagellates, 32-33
dragon fish, 52
Dubois, Raphael, 21

## E

Elateride, 28
enzyme, 21
esca, 50

## F

firefly, 18, 27-31, 63-64, 78-85
fireworm, 65
flashlight fish, 16, 49, 60-61
foxfire, 55
fungi, 16, 54-46

## G

glowworms, 14, 18-19, 59
Goddard Space Flight Center, 69

## H

Harrington, Anne, 50
Harvey, Dr. E. Newton, 22
Hastings, Dr. J. Woodland, 49
hatchet fish, 45
heart attacks, 72
hydroid. See Obelia

## M

McElroy, Dr. William, 23-24
medicine, 17, 67-73
metabolic energy, 16
mollusks, 41
moonlight mushroom. *See*
    Lampteromyces japonicus
Morin, James G., 50
muscular dystrophy, 72-73
mushroom, 16, 56
mycelia, 55
Mycena, 56

## N

New Zealand, 59

## O

Obelia, 39
oxygen, 19

## P

peeneewallies, 27
Photinus pyralis, 30

photometer, 49, 70, 76
photopores, 40
photoprotein, 38-39
Phrixothrix, 66
Picciolo, Dr. Grace, 69-70, 75
pony fish, 49
prey, 57-60
Princeton University, 22, 38
Project Firefly, 75-76

## R

railroad worms. *See*
    Phrixothrix
red tide, 33
Renilla, 39
Royal Society, 13

## S

sea fire, 33
sea firefly, 33
sea pansy. *See* Renilla
shrimp, 16, 52, 63
spiny dogfish, 48

## PICTURE CREDITS

American Museum of Natural History,
 pp. 51, 60, 62
Dr. Milton J. Cormier and James Anderson,
 p. 39
Y. Hameda (Princeton Univ. Biology Dept.),
 pp. 20, 41, 42, 45, 46, 56 left and right, 61
Dr. Frank Johnson, pp. 35, 36
William McElroy, p. 55 upper and lower
Bob Munns, U.S. Fish and Wildlife Service,
 pp. 34, 86
NASA, pp. 25, 71, 74
New York Public Library Picture Collection,
 p. 12
New York Zoological Society, p. 44
New Zealand Consulate General, New York,
 pp. 58, 59
Roy Pinney, p. 68
O. Shimomura (Princeton Univ. Biology Dept.),
 p. 37